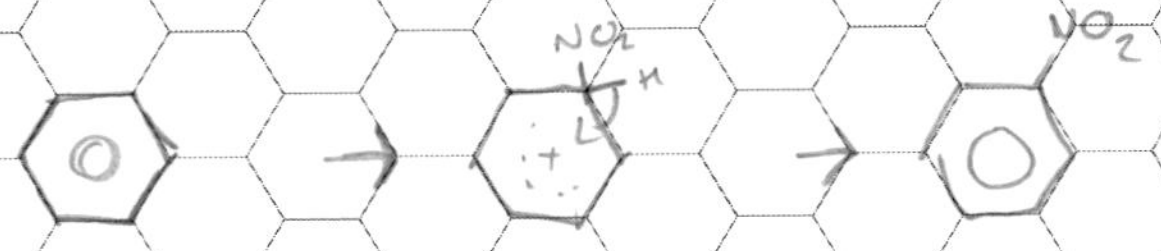
NO2
H
NO2

26824856R00086

Printed in Great Britain
by Amazon